Nuclear New Build?

Contents

Nuclear new build?

A review of the issues by
Christopher Gifford

Introduction

In June 2006, Spokesman Books published my pamphlet *Nuclear Reactors – Do We Need More?* in their Socialist Renewal Fifth Series, No 4. The pamphlet was a response to the last government's emerging policy of 'facilitating' the building of new nuclear power stations. That policy has now been in place nearly five years, but no application for a licence to build and operate a reactor has yet been made by any one of the two 'requesting parties' remaining in the 'Generic Design Assessment' (GDA) process – a process intended to 'fast track' design approval.

This paper attempts to describe the present state of affairs, in particular the issues facing the new Coalition Government. In February 2007, the then government was described in judicial review as having behaved 'unlawfully' in consulting on energy policy with information 'wholly insufficient for the public to make an intelligent response'. Since then, thousands of pages have been published in further consultations and some in response to freedom of information requests, and it has become clear that much detail remains to be provided on matters that may not be decided until licences to build and operate nuclear stations are granted, if at all. The material is usually technical, but there are ethical issues which demand political decisions after the involvement of an informed public. Meaningful information has been slow to emerge, and it is not surprising that, so far, few members of the public have become involved.

The requesting parties are the Westinghouse Electric Company, who submitted incomplete designs for an AP1000 Nuclear Reactor, and the Areva NP SAS and Electricité de France consortium, which offered designs for a European Pressurised Reactor – designs which were also found to be unsatisfactory, of which more later. Two other requesting parties, Atomic Energy of Canada Ltd and GE-Hitachi Nuclear Energy International, withdrew from the process.

Will the lights go out?

It has been suggested, often by those with an interest in nuclear reactors, that we are in danger of running out of electricity generating capacity if we do not quickly build nuclear generating stations. Coal stations, because of the need to limit CO_2 emissions, and nuclear stations nearing the end of their lives certainly will have to be replaced. Renewable energy projects, particularly tidal energy, have not been pursued with the urgency envisaged in the 2002 energy review, probably because of the then government's emphasis on nuclear. It has become clearer in the last year that the promise of proven, safe, efficient, economical 'generation III' nuclear reactors cannot be fulfilled, simply because they do not exist. It appears that, if they are permitted and developed, highly active spent fuel waste will have to be guarded in surface stores for more than 100 years – long after the companies producing the waste cease to have an income or even to exist. Our grandchildren's children would have to encapsulate the waste when it is cool enough and make it safe if they can.

The prospect of a nuclear renaissance resulted in a closer look at the industry's supply chain at a conference, 'Building a Nuclear Future', at the National Metals Technology Centre in Leeds in June 2010. A report in *Materials World,* the journal of the Institute of Materials, Minerals and Mining, states that keeping the lights on was a common theme but that companies were frank about the 'stark decisions' ahead. Dr David Powell, Vice President of Westinghouse UK, said 'The UK government needs to be committed to nuclear, we need low carbon generation as EU Directives (on carbon reduction) kick in'. He added that this move towards nuclear generation 'should have happened five to ten years ago'. Perhaps he should have mentioned that it did begin more than five years ago with the support of Tony Blair's government[1], and that not only was the industry unable then to offer the 'modern' reactors it claimed to have, but also, five years on, it has not yet offered designs acceptable to the regulators.

An understanding of the need to overcome the effects of a 20 year gap in nuclear reactor building led Lord Mandelson, then Business Secretary in the last government, to provide an £80m loan to Sheffield Forgemasters to design and build a 15 000 tonne press. Because only one similar press capable of forging large components for nuclear power stations existed worldwide, the Business Secretary saw the need to create such a facility in Britain. Both the former and the present government undertook that 'new build' should be without public subsidy and, without publication of the terms of the loan, there was doubt that, as with insurance waivers and dubious provision for waste management, the undertaking was being breached. On 17 June 2010, the new **Chief Secretary to the Treasury, Danny Alexander,** announced the cancellation of 12 projects totalling £2 billion agreed by the previous government, including the £80 million loan to Sheffield Forgemasters.

Similar evidence of ambivalence in the Coalition Government's position was the statement by **Secretary of State for Energy and Climate Change, Chris Huhne,** that a £4 billion 'black hole' existed in the finances of the Nuclear Decommissioning Authority (NDA)[2]. That should hardly be a surprise when a prervious Secretary of State had, in 2001, told parliament of a nuclear decommissioning legacy of £85bn[3]. But the new Secretary of State had noted that while government grants and income from Magnox electricity generation had kept the current Nuclear Decommissioning Agency budget in balance, annual deficits, that is, extra grants, of more than £1bn per year were foreseeable, without new build, when the last of the Magnox stations close and their decommissioning costs are added to the legacy.

Likewise, on 15 July 2010, the **Minister of State for Energy, Charles Hendry MP**, informed parliament that a 're-consultation' was to take place on draft Energy National Policy Statements to allow a fresh look at Appraisals of Sustainability. This would delay for perhaps a year the publication of any list of approved sites for reactor new build, but, amazingly, 'plans for the first new nuclear power stations to begin generating electricity by 2018 remain on course'. On the same day the Minister of State for Communities and Local Government announced a new streamlined system to replace the Infrastructure Planning Commission with an updated timetable.

Also in July the government-funded Offshore Valuation Group reported a practical resource of 2131TWh/yr (six times current UK electricity demand) of offshore renewable energy from wind, tidal stream and tidal range generators. Major expansion of the supply chain will be required to exploit the resource.[4]

There is no justification for talk of power cuts even though there are long lead times for both new nuclear and marine energy. More gas stations with combined heat and power production twice as efficient as existing coal or nuclear stations are already planned, and more will be needed. We have already benefitted from the lower carbon footprint of such stations. It is also possible to produce renewable gas from waste in greater quantities than we do now.

In April 2010, the then government announced a feed-in tariff scheme of up to 41.3p/kWh for local solar photo-voltaic electricity generation for domestic and other small scale consumers to generate their own electricity and feed back any surplus to the grid for installations completed by March 2012. Other renewable generation will be supported at lower rates.

The Conservative Lib-Dem Coalition

The previous government stated a policy of generating nuclear electricity 'without cost to the taxpayer'. The present Coalition Government has made a similar commitment in the Coalition Agreement for Government[5] which, on energy policy, contains many proposals for green and low carbon policies. The Lib-Dem part of the coalition has reserved the right to continue to oppose nuclear new build, which in the Energy and Climate Change section of the agreement appears as below.

> *Liberal Democrats have long opposed any new nuclear construction. Conservatives, by contrast, are committed to allowing the replacement of existing nuclear power stations provided that they are subject to the normal planning process for major projects (under a new National Planning Statement), and also provided that they receive no public subsidy.*
>
> *We will implement a process allowing the Liberal Democrats to maintain their opposition to nuclear power while permitting the Government to bring forward the National Planning Statement for ratification by Parliament so that new nuclear construction becomes possible. This process will involve: – the Government completing the drafting of a national planning statement and putting it before Parliament; – specific agreement that a Liberal Democrat spokesperson will speak against the Planning Statement, but that Liberal Democrat MPs will abstain; and – clarity that this will not be regarded as an issue of confidence.*

'No public subsidy' and 'No cost to the taxpayer' are clear commitments which can be construed as protection for future taypayers in the management of long-lived, highly active radioactive waste. Earlier proposals by the last government for such protection were far from convincing, and a draft Conservative 'Justification' Statutory Instrument has yet to be published for the obligatory consultation. There is further discussion in the section of this paper with the title 'Waste Management'.

Pollution

The Royal Commission on Environmental Pollution
Here it is necessary to record that the Coalition Government intends to abolish the Royal Commission on Environmental Pollution along with other quangos. No one has persuasively argued that the Commission's

principal 1976 recommendation, quoted below, is unsound. Its disregard is history. Safe management has not yet been demonstrated.

> *Some 30 years after the start of environmental pollution by radioactive waste on an industrial scale largely as a result of top secret work on atomic weapons the Royal Commission on Environmental Pollution in its sixth report[6] recommended that there should be no commitment to a large nuclear power programme 'until it has been established beyond reasonable doubt that a method exists to ensure the safe containment of long lived highly radioactive waste for the indefinite future'. Of the 'indefinite future' the Commission said*
>
> *We must assume that these wastes will remain dangerous and will need to be isolated from the biosphere for hundreds of thousands of years. In considering arrangements for dealing with such waste man is faced with time scales that transcend his experience.*

We note that the Commission in using the words 'beyond reasonable doubt'[7] chose a standard somewhat lower than certainty but, in spite of that and after thirty four years, we are still waiting for the implementation of the Commission's recommendation. A recent government statement concedes that safe containment for an indefinite period will never exist. In the consultative document *Managing Radioactive Waste Safety – a framework for implementing geological disposal*, 25 June 2007, the Department of Environment stated[8]

> *It is inevitable that some radioactivity will eventually reach the surface.*

followed by

> *But the disposal facility is designed to ensure that this will not happen for many thousands of years, and even then only in quantities that are insignificant compared to the levels of radioactivity all around us in the environment from natural background sources.*

The present tense of 'is designed', rather than the alternative 'will be designed' used in a later document, conveys the hesitation and the wishful thinking in what would otherwise be a strongly affirmative statement. The evidence for the affirmation is not provided in either document. The reader was referred to the recommendations of the Committee on Radioactive Waste Management (CoRWM) in Chapter 13 of the Committee's Final Report[9].

The Committee's Reservations
Chapter 13 describes 'The wider science community's views and challenges to those views', before stating the Committee's unanimous

support for its recommendation as stated in Chapter 14. Recommendations 1 and 4 are quoted below to illustrate the less affirmative position and the reservation that much research is still required to reduce uncertainties.

> *Recommendation 1 – Within the present state of knowledge, CoRWM considers geological disposal to be the best available approach for the long-term management of all the material categorised as waste in the CoRWM inventory when compared with the risks associated with other methods of management. The aim should be to proceed to disposal as soon as practicable, consistent with developing and maintaining public and stakeholder confidence.*
>
> *Recommendation 4 – There should be a commitment to an intensified programme of research and development into the long-term safety of geological disposal aimed at reducing uncertainties at generic and site specific levels, as well as into improved means for storing waste in the longer term.*

The Committee made it very clear that its terms of reference were to advise on the treatment of legacy waste and that any question of building new nuclear reactors was beyond its brief. The recommendations quoted above contain the nuance that geological disposal is the option likely to cause least harm rather than that it is the 'solution' mentioned in a later white paper which is discussed below. Since then a re-constituted CoRWM has been formed.[10]

History

A History of Mendacity

Those of us old enough to recall government and nuclear industry statements over the years that the industry is economic, peaceful, safe and necessary have cause to be sceptical of current proposals. That electrical power generation in civil nuclear reactors was a peaceful activity was a fiction maintained for more than 20 years. Worse than that was the assertion, maintained until 1978, that spent fuel had not been used to make nuclear weapons and *could not be so used*. Even as late as 1989 an acknowledgement that spent fuel from civil reactors had been used to make nuclear weapons was disputed by the Central Electricity Generating Board (CEGB) representatives at the Hinkley Point 'C' Public Inquiry, but a quotation from the late Lord Hinton, a former chairman of the CEGB, was offered in evidence. **He was reported to have said in response to the statement that 'No plutonium produced in CEGB reactors has been applied to military use either in the UK or elsewhere'**

I am absolutely certain that that statement is incorrect. I am questioning the whole statement because it is deplorable … What is important is that they shouldn't tell bloody lies in their evidence.[11]

Sir Walter Marshall, later Lord Marshall, was also quoted by the same witness as having written to *The Times* which, on 6 June 1986, published his letter containing

I said that plutonium produced in the early years of operation (of) the first nuclear stations had gone into the defence stockpile.

Today the fact that civil nuclear reactors produce plutonium which can be processed into nuclear weapons is acknowledged beyond dispute. It is stated as the concern that Iran may produce such weapons if it continues lawfully with a civil nuclear power programme.

The last government proposed to dispense with public inquiries and we wait to hear what a Conservative 'Justification' statement will include. We need to remember that one of the essential functions of a public inquiry is to examine false or misleading claims to establish the truth; and that at our last public inquiry into a proposed nuclear power station that is exactly what happened.

Nuclear reactors would be peaceful if there were means of ensuring that spent fuel would not be processed into weapons. Means such as safe disposal or effective international control do not exist. It is the opinion of Professor Fred Roberts, a former UK Atomic Energy Authority researcher, in *Sixty Years of Nuclear History*[12], that effective nuclear disarmament will not be achieved while nuclear reactors exist.

There are many technical, ethical, political and environmental judgements to be made about nuclear power generation. According to Sir James Lovelock, author of the Gaia hypothesis and one-time green environmentalist, public opinion is set against nuclear power because of 'ceaseless misinformation from the green lobbies'[13]. After a visit to advocate nuclear power generation in response to global warming, he described Sellafield as 'clean and tidy' and failed to mention that the Nuclear Installations Inspectorate (NII) had described the site as having some features too dangerous to examine; leaking tanks, insecure and overloaded structures, and unconditioned plutonium-containing waste vulnerable to criticality.[14] He also failed to mention that unauthorised as well as authorised discharges to the sea contaminated 40km of beaches and led to the exclusion of the public and the successful prosecution of the company, BNFL. The Irish Sea remains the most radioactively contaminated marine environment in the world. Many foreign governments, dozens of local authorities and thousands of individuals have objected. There are clusters of childhood leukaemia. Children throughout Britain have plutonium in their teeth and bones.[15] An added irony for Sir James is that it was Greenpeace activists who brought the unauthorised discharges to public attention.

The 2002 Energy Review

In 2002, the government published an Energy Review[16]. In over 200 pages of detail it discussed options for future supplies of energy. It was written by the Performance and Innovation Unit of the Cabinet Office, but it has since become clear that the Department of Trade and Industry, although involved, was not the principal author.

On the generation of electricity in nuclear power stations, the review said that concern about radioactive waste and 'low probability but high consequence hazards' may limit or preclude its use. It added that nuclear power seemed likely to remain more expensive than fossil-fuelled generation and that nowhere in the world was there new build in a liberalised electricity market. Thus two of the objections of those opposed to nuclear power were conceded. It was not safe and it was not economic. Similarly, the report mentioned the vulnerability to terrorism, the long lead times in planning and building new stations, the extent of public opposition, and the need to gain public acceptance for any new development. It noted that the nuclear waste issue was unresolved. It concluded that the option of new investment in nuclear power should be kept open, especially if safer and low-cost designs were developed, but there would have to be widespread public acceptance.

A major stakeholder and public consultation was launched in May 2002. It was the largest ever on energy policy. There followed a white paper which concluded that diversity of supply was the best protection against sudden price increases, terrorism and other threats to reliability of supply.

On renewable energy, the review had concluded that 'the UK resource is, in principle, more than sufficient to meet the UK's energy needs' and that 'the UK's wind and marine resources are the best in Europe'. Both publications were strongly focused on the need to mitigate climate change. The review had already stated that, while achieving a 60% cut in CO_2 emissions would be challenging, it could be done while still achieving economic growth of 2.25% per year.

The Review reviewed

It did not make sense that global warming and security of supply should be cited as the reasons for another energy review in 2005[17]. But that is what happened, and the prime minister, Tony Blair, who had written the preface to the first review and endorsed the detailed conclusions on those matters, declared that the building of new nuclear power stations should be 'facilitated' by 'fast track' planning inquiries and 'pre-licensing' of new reactor designs. Another public consultation followed.

This writer responded to these events with some dismay, and wrote a paper with the title *Nuclear Reactors: do we need more?*. The paper was published by the Bertrand Russell Peace Foundation in the Socialist Renewal series and a review and an abstract appeared in *The Spokesman* journal.[18] It examined the historic claims that nuclear power was peaceful and safe and asked 'Is the risk from terrorism too awful to be acknowledged?' It described the failure in the UK to comply with a European directive on the provision of information to the public on possible emergencies, examined the lack of data on costs, discussed the known costs but lack of solutions on nuclear waste management and listed the, so far, neglected sources of safe, sustainable renewable energy.

The response of the Nuclear Installations Inspectorate to the government's proposals was reassuring. The Health and Safety Executive (HSE) endorsed the concerns of the Nuclear Safety Directorate by publishing a 150 page expert report with the title *The Health and Safety Risks and Regulatory Strategy Related to Energy Developments*[19], which emphasised the importance of the licensing process to control risk by the design of licence conditions after detailed appraisal of a reactor design and the builder's safety case. The HSE made no concessions to the prime minister's proposals. It explained that if the (13) vacancies for government inspectors were filled quickly the study of a designer's safety case and proposed reactor for a specific location would take several years (as it always had) depending on the quality of the application. If more than one new design had to be appraised concurrently it would take longer. The publication reported on earlier experience of 'pre-licensing' and mentioned the Commission's finding in a 1994 review that the regulatory systems were 'comprehensive, internationally recognised, vindicated by public inquiries, and that there was no reason to change them in any fundamental way to deal with changes to the nuclear industry or new construction'.

It is difficult to imagine a more severe reprimand of a lay prime minister's interference in a process vital to public safety. Public concern about the government's methods was not alleviated by the HSE response. Greenpeace, with the support of other organisations such as the Welsh Anti-nuclear Alliance (WANA) and the Nuclear Free Local Authorities, applied to the High Court for judicial review of the way in which the government had consulted the public while giving every indication of having already decided the matter.

Here it is necessary to make comparison with Mr Blair's treatment of Iraq's supposed weapons of mass destruction. There are common ingredients such as a culture of compliance with the wishes of a prime minister who, it was later found, could be 'free from doubt' when he had been advised otherwise. We know that there was a dearth of meaningful debate in the cabinet itself.

A 'seriously flawed' and 'unlawful' consultation

Mr Justice Sullivan in the High Court on 15 February 2007 ruled that the government's second consultation on energy policy was 'seriously flawed' and thus 'unlawful'. There had been no consultation at all, he said, because the government had provided information 'wholly insufficient for the public to make an intelligent response'. In fact the government had also blacked out the economic data in papers obtained by the provisions of the Freedom of Information Act.

The government was obliged to start again. It published two white papers, one, *Planning for a Sustainable Future*[20], dealing with planning procedures, and *The Energy White Paper*[21], which was linked with a consultative document on nuclear power[22]. The documents, like the process criticised in judicial review, showed the government's commitment to nuclear power, which, this time, was described as a 'preliminary view'. The energy white paper is 343 pages long and is characterised by enhanced optimism and a lack of vital facts. I tried hard to find, for example, data on the present and historic costs of generating electricity by nuclear power, but I found none. Instead there are unattributed forecasts of future costs only one of which favours nuclear power – that which assumes high gas prices and generous carbon credits.

Generic Design Assessment

The 'pre-licensing' process, now renamed the Generic Design Assessment (GDA) process, was intended to 'fast-track' the licensing process. There is no indication in the reports of the Nuclear Installations Inspectorate and the Environment Agency (EA) that any faster process has been adopted. The 'requesting parties' (the vendors) may gain for prospective operators and the public some limited reassurance from early non-site specific assessment of the designs if sufficient information is provided. The early indications are that more intelligible information is needed.[23] It is not clear at what stage a requesting party is to be charged for the generic assessment but NII commentary has indicated that all assessment must be paid for. The Generic Design Assessment process is not obligatory.

Eventually applications from prospective operators for site specific Site Operating Licences have to be made and determined by the Nuclear Installations Inspectorate. There is no report to date that any application has been made.

The Nuclear Safety Directorate (now renamed 'The Nuclear Directorate') reported in their e-mail bulletin dated 4 August 2008 on the progress of the Generic Design Assessment of new nuclear reactor designs, and there have been regular reports since. HSE and the Environment Agency were involved in the initial assessment of four designs. These were

- The Areva NP SAS and Electricite de France SA UK EPR Nuclear Reactor
- The Atomic Energy of Canada Ltd ACR-1000 Nuclear Reactor
- The GE-Hitachi Nuclear Energy International LLC ESBWR Nuclear Reactor and
- The Westinghouse Electric Company LLC AP1000 Nuclear Reactor.

The interim reports were to the effect that no obstacle to further assessment had been found.

Since then the designs for the ACR-1000 reactor and the GE-Hitachi Nuclear Energy International LLC ESBWR (Economic and Simplified Boiling Water Reactor) have been withdrawn from the Generic Design Assessment process.

Safety

'Safety is not an issue'
During a 2006 House of Commons exchange on nuclear power the then new leader of the opposition, David Cameron, declared that safety was no longer an issue, and the grateful then prime minister,

Tony Blair, even with his better information about the vulnerability to terrorism, took no exception to that claim. Such a 'consensus' is dangerous and suggests that those involved believe the recent 'spin'.

In the consultative document, *The Future of Nuclear Power*,[24] there is frankness combined with optimism in the discussion of the dangers of nuclear power, as in

> 'Not all costs are considered. The analysis does not attempt to monetise all costs and benefits. Specifically, a monetary value associated with potential accidents is not estimated. Evidence suggests that the likelihood of such accidents is negligible, particularly in the UK context.'

The justification for the above is found in a footnote which reads

> The literature suggests a range for the probability of major accidents (core meltdown plus containment failure) from $2x10^{-6}$ in France, to $4x10^{-9}$ in the UK. The associated expected cost is estimated to be of the order £0.03 / MWh to £0.30 / MWh depending on assumptions about discount rates and the value of life; using the figure at the top end of this range would not change the results of the cost benefit analysis. Introducing risk aversion, the results of the cost benefit analysis in the central case (defined in Section 3 below) would be robust for a risk aversion factor of 20 at the highest estimated value for the expected accident cost. For a summary of the relevant literature, see 'Externalities of Energy (ExternE), Methodology 2005 Update', European Commission.

The consultative document quoted here contains contradictory information about the probability of loss of containment of nuclear reactors. The claim that the risk of meltdown and loss of containment of a reactor is 'negligible' was based on two different probabilities attributed without authorship to the European Commission. At page 66 the probability is stated as 4×10^{-9} and at page 105 is stated as 'one in 2.4 billion per reactor per year'. (Probability here is a number less than one. Expected frequency is the reciprocal of probability.) The latter estimate resolves to a probability of 4.2×10^{-10}. They differ by an order of magnitude, but they are incredible for other reasons. At the last public inquiry the revised estimate produced by the Director General of the Health and Safety Executive was 1 in 100 000 per reactor per year, a probability of 1×10^{-5} (revised from 1×10^{-6}). It was challenged as being no more than a guess.[25] We are now being asked to accept that reactors are 24 thousand times safer than in 1989, without being told the name of the author of the estimate or any of his or her reasons. And the debate on whether or not our nuclear reactors are capable of *nuclear* explosion is not yet settled.[26] The nuclear industry remains uninsurable.

These contradictory and incredible statements confirm the impression that the second consultation was as inadequate as the first, which was rejected as misleading and unlawful in judicial review. They probably appeared because Tony Blair could not find anyone better informed to write about risk assessment. The writer chosen did not know his or her mantissa (better called a 'significand') from his or her exponent; or, as they say in Yorkshire, didn't know his arse from his elbow. The writer was probably chosen for the ability to 'sex up' a document. But what is happening when people, clearly not capable of dealing with the topic, are instructed to write in support of a policy which has already been challenged by the judiciary as poorly presented and ill informed? Was there discussion in cabinet when the policy was switched from renewable energy to nuclear new build?

On 3 June 2009, I sent an e-mail to Adam Dawson at the Department for Energy and Climate Change (DECC) offering my paper *Geological Disposal of Nuclear Waste*[27], which presented arguments similar to those stated here proposing an independent inquiry into new build rather than a Justification decision by a member of a government already committed to it. I asked if the Department intended to publish a correction of the transparent mathematical error shown above. I received no reply.

The estimates above must include the probability of impact by aircraft and missiles, unless reactor meltdown by hostile process has been excluded. If that is the case, it is a qualification well worth mentioning. In a parallel consultation, discussing verification of nuclear power station designs to withstand impact by a 590 tonne aircraft flying at 550mph, the Health and Safety Executive at the Nuclear Installations Inspectorate promised a reply by 17 October 2007. None has been received. I had already noted that documents supporting the adequacy of the design were restricted, but not even the formula used to assess an ability to withstand impact could be supplied. The Environment Agency promised me a reply by 24 October 2007, but none has been received.

Those in any doubt about whether safety is an issue need only look at Regulations 14 and 18 of the Radiation (Emergency Preparedness and Public Information) Regulations.[28] Regulation 18(2) empowers the Secretary of State for Defence to exempt Her Majesty's forces and others involved in defence from all or any of the regulations. He is separately empowered by Regulation 18(3) to direct verbally that the requirements of Regulation 14 shall have no effect to the extent that this regulation would, in his opinion, be against the interests of national security. Regulation 14 deals with exposure to radiation in emergencies and includes requirements on training and the provision of equipment and information to those who may be permitted by an authorised person to be exposed to exceptional prescribed doses of radiation. A volunteer informed of the risks may agree to an unlimited dose for the purpose of saving life.

That the government should envisage it necessary to waive such accommodating provisions is chilling. It is now deemed necessary to be ready to set aside the need for volunteers to be told the extent to which their lives may be at risk, or for persons who may not be volunteers to be instructed by persons lacking authority to instruct.

'Safe and secure' – say it often

It has been quietly acknowledged in recent publications such as the *Draft National Policy Statement EN-6*[29] that spent fuel waste will be stored at the surface of nuclear power stations for many decades, and that 160 years could elapse before some of the waste from new build could be placed in a geological depository, because such spent fuel from high burn-up reactors will produce more heat for longer periods than does legacy waste. We are asked to accept that highly active spent fuel will for the same periods of time be stored 'safely and securely'.

The words 'safe and secure' or 'safely and securely' occur 25 times in this consultative document, and 40 times in the draft National Policy Statement EN-6.[30] It is as if someone thought that repetition would make it more convincing. The impression given is one of unwarranted optimism in statements that often lack supporting evidence, or have evidence cited that lacks an author's name. The standard required in documents of this importance is that which would stand up to rigorous peer review by experts not already committed to the expansion of the nuclear industry.

How can surface storage be safe and secure given that terrorists have already demonstrated aerial impact on an appalling scale? In a meeting with six Nuclear Installations Inspectors on 22 June 2007, convened to discuss graphite core degradation in Magnox reactors, I asked for attribution of the opinion that no enlarged emergency planning zone need be advised to members of the public with information on their evacuation, treatment, transport and shelter.[31]

The inspectorate had been advised not to discuss such matters by the Office for Civil Nuclear Security (OCNS). But the OCNS is part of the Nuclear Directorate. The Nuclear Directorate had removed itself from anything resembling peer review on this issue. For the present it seems that 'security' demands that the law be ignored, that people shall not be told when they are at risk, and that those described as responsible for those decisions shall not discuss them.

Waste management

The government claimed in the 2006 consultative document[32]

> *We have technical solutions for waste disposal that scientific consensus and experience from abroad suggest could accommodate all types of waste from existing and new power stations.*

Here the word 'disposal' has displaced the earlier mention of a 'repository', and 'depository' and 'repository' now have the same definition as a place for disposal.[33] The findings of the Committee on Radioactive Waste Management have been misreported to turn a topic requiring further research and a suitable site into a solution. Waste from the recently encouraged new build, which CoRWM expressly excluded from its considerations, is to be included in a 'co-disposal' depository for legacy waste and highly active spent fuel waste from new reactors.

Advice from the International Atomic Energy Agency (IAEA)
The mention of 'all types of waste' here includes the spent fuel after decades of storage from the proposed new nuclear power stations, which it is proposed be operated at higher burn-up rates. The US Nuclear Regulatory Commission has expressed concern about higher burn-up rates.

> *... there is limited data to show that the cladding of spent fuel with burnups greater than 45,000 MWd/MTU will remain undamaged during the licensing period. Limited information suggests increased cladding oxidation, increased hoop stresses and changes to fuel pellet integrity with increasing burnup up to and beyond 60,000 MWd/MTU. These burnup dependent effects could potentially lead to failure of the cladding and dispersal of the fuel during transfer and handling operations.*[34]

Safety fears about the longer term integrity of such fuel is becoming an international matter, leading the International Atomic Energy

Agency (IAEA) to demand more research on fuel behaviour in dry storage as essential. The Agency advised

> *'In particular ... high burnup fuels and mixed oxide (MOX) fuels will need to be carefully assessed in the context of ensuring long term storage safety.'*[35]

The proposal for underground co-disposal of legacy waste with waste from new build has to be appraised against the facts that no such disposal facility exists anywhere in the world, that no site for a geological disposal facility has yet been found in the UK, that the Nuclear Decommissioning Agency (NDA), who are to provide the facility, is already postponing work for lack of funds, that further cuts are planned, that none is expected to be built before 2040, and that it is already conceded, as noted earlier, that it is inevitable that some radioactivity will eventually reach the surface.[36]

That there are technical solutions is firmly contradicted in the Committee on Radioactive Waste Management Document 2500, *Outline of CoRWM Interim Storage Report March 2009*, which in paragraph 3.5 states

> *However, it is our unanimous opinion that greater attention should be given to the current management of radioactive waste held in the UK, in the context of its vulnerability to terrorist attacks. We are not aware of any UK government programme that is addressing this issue with adequate detail or priority, and consider it unacceptable for some vulnerable waste forms, such as spent fuel, to remain in their current condition and mode of storage.*

The careful reader of the government's optimistic claims will note that the particular claim that there are technical solutions was made in May 2009, two months after the government received the updated summary of the advice quoted above. Vulnerability to terrorism was not the only concern of the Committee on Radioactive Waste Management. In the same document there are many others. In paragraph 3.6, for example:

> *In the case of radioactive wastes destined for geological disposal, transport will take place decades after the wastes have been conditioned and packaged. There can be no guarantee that a waste package designed for transport now will be suitable after decades in store.*

As we have seen, the US Nuclear Regulatory Commission has expressed concern about higher burn-up rates. The National Policy

Statement EN-6[38] maintains the hubris with

> *Having considered this issue, the government is satisfied that effective arrangements will exist to manage and dispose of the waste that will be produced from new nuclear power stations. As a result the IPC (Infrastructure Planning Commission) need not consider this question.*

The public inquiry option should be exercised
That last sentence signals the end of scrutiny by any public planning inquiry and the exclusion of the public from further involvement in national planning if a Statutory Instrument is made as drafted. The option of holding a public inquiry into a draft decision on justification remains with the Secretary of State and it should be taken. That and much more will be required to meet the High Court's standard of a meaningful consultation and before the public can be expected to share the government's so easily found satisfaction that effective arrangements will exist in 160 years time.

Misinformation by DEFRA
This paper is inevitably a commentary on the contradictory information on nuclear industry affairs. On nuclear waste management similar contradictions exist. In September 2001, the Department of Environment, Food and Rural Affairs (DEFRA) published *Managing Radioactive Waste Safely – proposals for Developing a Policy for Managing Solid Radioactive Waste in the UK*.[39] It was the first of a series of documents with that title and it began with a statement in the executive summary:

> *More than 10 000 tonnes of radioactive waste are safely stored in the UK, but await a decision on their long-term future.*

My response to the consultation included

> *The DEFRA report is misleading and inaccurate in its opening statement – that waste is presently managed safely. The statement is dangerously complacent and disregards much that the Nuclear Installations Inspectorate has published recently. The report should be recalled and corrected. It should also be edited to refer to the present problems of unsafe storage, the recommendation of the Royal Commission on Environmental Pollution, the other consequences of radiological pollution such as the mutagenic effects and the loss of land and habitation, and the present estimate of a clean up cost of £85 billion if no expansion of the industry occurs.*

In *Safety Audit at Dounreay 1998*[40] the Nuclear Installations Inspectorate (NII) reported that the UK Atomic Energy Authority (UKAEA)

> *used the Dounreay Shaft (D1225) for disposal of solid waste between 1959 and 1971. In 1971 the Wet Silo came into service as an intermediate level waste store. The shaft was used until 1977 for items that were too large for the Wet Silo (D9833) when an explosion in the shaft led to a cessation of input of material. There is considerable uncertainty over the contents of the shaft, but it is believed to contain equipment contaminated with radioactive material and sodium, chemicals, natural uranium fuel, radioactive sources, incinerator ash, filters, gloveboxes, building materials, sludges, clothing etc. ... UKAEA accepts that the shaft does not meet current standards for an intermediate level waste disposal facility. The Government has recently accepted that UKAEA's proposal to retrieve the waste ... The plan is to carry out the work between 2014 and 2018.*

The explosion on 10th May 1977 was probably caused by sodium reacting with water to produce hydrogen. The explosion blew off the 12 tonne cap at the top of the shaft and created hundreds of hotspots in the area and along the coastline. The effects of the explosion were not reported to the Committee studying the Medical Effects of Radiation and investigating leukaemia clusters in the United Kingdom.

The Nuclear Installations Inspectorate also reported that the waste in the Silo was

> *not in a safe passive form.*

After interviews with UKAEA officials, John Aldridge writing in *The Guardian* on February 2nd 1998 reported that over 1000 tonnes of waste was to be removed from the shaft and some 700 tonnes of waste from the Wet Silo, and that there were fears that another explosion could occur. Morris Grant, a spokesman for the Authority, was quoted as saying that the clean up using ground freezing techniques and robots would cost up to £1 billion. On 20 February 2007, the start of a project to isolate the shaft by rock grouting was described to members of the Institute of Materials, Minerals and Mining by David Gibson of the Ritchies Division of Edmund Nuttall Ltd.[41]

The Nuclear Installations Inspectorate reported other examples of unsafe storage at Dounreay and at many other sites in the UK. In 1999, the NII reported that some of the liquid waste at Dounreay from reprocessing was stored in 15 stainless steel tanks, some dating back to the 1950s. The report explained that inspection was not easy but that measurements indicated that the tank walls were still in good condition. There were, however, signs of deterioration in the wall of the cells housing the tanks. The work of encapsulating the intermediate level waste raffinates will take until 2012 to complete. There follows a qualified statement of little comfort to the Government which recently provided the NII with this under-regulated legacy of neglect and mismanagement

We consider that the position with the raffinates is adequate provided that there are no delays in the cementation programme, and no further deterioration in the storage facilities. The flocs are stored in three tanks. One of the older tanks has developed a small leak.

The Nuclear Installations Inspectorate was not satisfied with the storage and treatment of high level waste either. In its report of February 2000 it said that it was 'unconvinced' that British Nuclear Fuels Ltd (BNFL) at Sellafield could clear the backlog of Highly Active Liquid (HAL) by an agreed date of 2015. On receipt of the plan for stock reduction to passive states, the NII issued a Specification enforceable as a Site Licence Condition to require improvement in the rate of vitrification and other constrains.

The DEFRA report was about solid, intermediate and low level radioactive waste, but even with those restrictions, which were made elsewhere than in the Executive Summary, the opening statement about safe storage was simply not justified. It is particularly offensive that it should appear without qualification in the 'Executive Summary'. Examples of plainly unsafe procedures were presented in the text, for example, at page 57, paragraph 7.8 where wastes are held in a raw untreated state:

these should be made passively safe … as soon as possible.

But they were unlikely to be read by those 'Executives' who trust others to provide summaries and who may go on to support action which will add to the already irremediable problems of a contaminated planet. The Executive Summary was comfortingly silent on the costs of managing the waste generated to date, but the narrative of the report included an estimate of £40bn as the cost of civil nuclear waste management and decommission. This did not include the cleaning up of military nuclear waste in the UK. On 18 October 2001, the Secretary of State, in reply to questions in the House of Commons, provided a total estimate, including military clean up, of £85 billions.

To clean up the US weapons sites alone will cost 200 billion dollars, according to Walter Stahel of the Geneva Association. 'Clean up' as used here means only that the material is stabilised and perhaps put somewhere else. The £5bn Sellafield 'Rock Characterisation Facility' project, with a view to creating an underground repository, was abandoned in 1977. No project in the UK has yet shown that safe burial is possible.

The treatment of plutonium as an asset with fuel making potential, when there is no customer in sight, amounts to dubious accounting on the basis of which BNFL once claimed a financially viable future. It is also cavalier and hardly consistent with the Nuclear Non-proliferation Treaty to create for profit an international trade in materials that can be turned into weapons. The recommendations of the Royal Society Report, *Management of Separated Plutonium*, the House of Commons Trade and Industry Select Committee report on the BNFL Public Private Partnership, and the House of Lords Select Committee report were in close agreement. For environmental and security reasons it would also make sense to have all nuclear weapons and nuclear material with weapon making potential demobilised and regulated much more severely than at present. The proposal to treat plutonium as waste was sound.

The government has yet to decide the matter, but *Managing Radioactive Waste Safely* (June 2008[42]) contains the statement that the current owners of depleted uranium and plutonium

> '... place a zero asset value on these radioactive materials meaning that they are neither classified as waste nor a commercial asset.'

When an organisation is succeeding in spite of an inattentive government that seems to be disputing or ignoring its problems, there is cause for concern. The government proposes, for example, more privatisation, the 'discipline of the private sector' and more use of contractors when privatisation and the delegation of core management functions to contractors have already been described as serious problems.

> *Our main finding is that organisational changes made within UKAEA ... have so weakened the management and technical base at Dounreay that it is not in a good position to tackle its principal mission ... The changes include the loss of experienced staff ... in the drive towards contractorisation.*

We now find that UKAEA is over-dependent on contractors for the delivery of many of the key functions which we would expect to see under the clear control of UKAEA as the licensee for the site.

We found that in many cases control of activities had been delegated too far, such that UKAEA was not in control, nor was it in a position to understand the safety significance of the contractors' activities. We believe that it is essential for UKAEA to re-establish effective control of nuclear safety related activities at the site.

DEFRA's misrepresentation of the state of nuclear waste was put to the Committee on Radioactive Waste Management in correspondence after an appearance at one of its meetings made accessible to members of the public in Cardiff. Two topics were raised by members of the Welsh Anti-Nuclear Alliance with immediate responses suggesting that at least some committee members had taken the issues on board. The first was the time scale that CoRWM was using to frame a recommendation on waste storage or deposition. The problem is much easier if one thinks of a few hundred years instead of the hundreds of thousands of years mentioned by the Royal Commission. The second topic was that of depleted uranium.

Depleted Uranium

Depleted uranium was first described as mainly the isotope Uranium 238. To make reactor fuel or an atomic bomb natural uranium has to be enriched by increasing the proportion of the isotope uranium 235 from 0.7% to more than 2%. The uranium 235 then will support a chain reaction with the release of much energy. The uranium metal from which the fissile isotope 235 has been extracted to make fuel for Magnox reactors is called depleted uranium.

Depleted uranium (DU) has a density of 18.9. It is toxic as well as being mildly radioactive with a half-life of 4.5 billion years. In spite of the toxicity and the ability to cause cancer and genetic mutations, the military found it useful to increase the penetrating power of shells and bullets, and even to improve the armour on military vehicles. DU munitions were test fired in Britain and the USA in the 1980s and used in Iraq in 1991, in Bosnia in 1996, in the Kosovo conflict in 1999, in Afghanistan in 2002, and in Iraq in 2003. It was estimated that the amount of DU used in the 1991 Gulf war was 340 tonnes. In the 2003 attack on Iraq up to 2000 tonnes may have been used, with up to 7 tonnes used in single 'bunker busting' bombs.

Servicemen and women's organisations and others interested in the health of service personnel and civilians questioned the consequences of battlefield exposure to radioactive and toxic materials inhaled as dust or ingested with food. The Ministry of Defence (MoD) response was unequivocal. The risks were negligible except for persons who remained for a long period in a vehicle hit by such a weapon, and the MoD denied the contrary findings of its own leaked report as 'a discredited draft prepared by a trainee'. But independent researchers took samples from service personnel indicating the ingestion of 15 times what the MoD had described as a 'safe dose'. Most physicists agree that there is no such thing as a safe dose. Scientists from the UN Environment Programme called for recoverable fragments of DU to be removed from conflict sites. The Royal Society also called for sampling, clean up and monitoring.[43]

In his book *Sixty Years of Nuclear History,* published in 1999, Fred Roberts, a former atom bomb scientist, described depleted uranium also as a product of the reprocessing of spent fuel from nuclear reactors. Within a few months, Paul Brown, the environment correspondent of *The Guardian,* after discussions with MoD staff but without attribution, also described DU as a product of reprocessing. The awful truth was out. The nuclear industry and the MoD had not only found a new way of dealing with mildly radioactive 'natural' nuclear waste. It was helping to dispose of waste from reactors and reprocessing plants which would contain transuranic elements, even allowing for the fact that at least some of the plutonium had been recovered.

Transuranic elements such as plutonium are formed in nuclear reactors and are not normally found in the earth's crust. When the UN environment programme found traces of plutonium and other highly radioactive particles in Kosovo, the MoD and the US Department of Energy admitted that the material came from depleted uranium shells, but denied that the uranium had been reprocessed. The uranium had been 'accidentally contaminated' in containers containing reprocessed materials. Two months later, the UN Environment Programme report on sites in Bosnia referred to 'huge variations' in plutonium levels in pieces of munitions found.

Explanations of 'accidental contamination' became unnecessary in November 2001. The UK Environment Agency commissioned and published a report, 'Depleted Uranium: a Study of its Uses within the UK and Disposal Issues'.[44] In a general description of depleted uranium the report states in an opening paragraph, 'Depleted uranium (DU) is the main by-product of the uranium enrichment process wherein the content of the fissile isotope U235 is enhanced in relation to the U238 content. In addition DU is produced from the reprocessing of Magnox reactor fuel in the UK'. A similar extended definition of DU appeared in September 2001, when the Department for Environment, Food and Rural Affairs published policy proposals for the management of radioactive waste.[45] DEFRA had little to say about the military use of uranium metal, but defined *depleted reprocessed uranium* as a sub-category of depleted uranium.

In response to the DEFRA proposals I wrote

Explanations are now needed on the accuracy with which other transuranic radioactive material is removed from spent fuel before it is released for use as munitions and by whose authority it is released. We are here discussing what to do with nuclear waste and learning, in passing, that firing it at one's enemies is a legitimate method of disposal! Such use should be prohibited by the UK government and by international agreement.

The Environment Agency report estimated worldwide stocks of DU at well over one million tonnes. The total is estimated to double by 2015. It is by no means the most troublesome of the nuclear industry's waste – plutonium is toxic, highly radioactive and an atomic bomb material.

The Ministry of Defence justifies the use of DU because to desist from its use would expose British service personnel to greater risks. There is no doubt that guided weapons, satellite technology and the greater penetrating power of bombs and shells were major factors in the military supremacy which led to the rapid defeat of Iraqi forces. But the use of toxic and radioactive materials is a form of chemical and nuclear warfare no different from the use of a radioactive 'dirty bomb' postulated as a possible terrorist weapon. The effects on the environment will last for thousands of years with many generations exposed to genetic effects. International agreement on the prohibition of such weapons and release of civil nuclear materials for military purposes is needed, and the countries best placed to bring that about are the United States and the United Kingdom.

The Committee on Radioactive Waste Management, when creating a short-list of options for waste management, did not include using spent fuel as weapons. That alone may not be sufficient to persuade a government to desist, but it might help to get it on a disarmament agenda. Some members of the Committee were slightly shocked to hear of the admissions, and one provided a Ministry of Defence source who had data on the low concentrations of the transuranic nasties in DU. Can we believe that all samples will be so clean?

The Nuclear Decommissioning Authority
The Nuclear Decommissioning Authority (NDA), incorporating the former NIREX (originally the Nuclear Industry Radioactive Waste Executive), was set up on 1 April 2005 under the provisions of the Energy Act 2004 and charged with the task of managing Britain's nuclear waste. That includes decommissioning and clean up at all civil public sector sites including 19 former BNFL and UKAEA sites. The sponsoring department was the then Department for Business, Enterprise and Regulatory Reform (DBERR), which approved the strategy, plans and budgets.

The budget for the year 2009-10 is £2.8bn of which £1.63bn was projected income (from electricity sales, reprocessing, land sales, etc) and the remainder government grant-in-aid. The expenditure for the last financial year is reported as £2.72bn. In previous years it has been reported that work on several projects was deferred for lack of funds. The nuclear liabilities have been reassessed upwards by £400m to £44.5bn. The annual report for 2008-9 states

> ... the estate that we inherited has proved to be more challenging than previously understood and demonstrates that we are in an evolving situation.

The Authority's web site states:

> We do not directly manage the sites for which we are responsible. Instead we contract out the delivery of site programmes through management and operation contracts with licensed operators, Site Licence Companies (SLCs), at each site.
>
> SLCs manage sites, including preparing site plans, performing and sub-contracting work. Parent Body Organisations (PBOs) own shares in SLCs for the duration of their contract with the NDA. The PBO is responsible for managing the delivery of site programmes. The contracts with PBOs are periodically competed.

Since the election the Coalition Government has sought cuts in the decommissioning budget. Members of the Prospect trade union representing Sellafield employees debated a deficit of £107m in the budget and envisaged that 800 employees would be made redundant.

The Nuclear Decommissioning Authority's finances are mainly to do with waste arising from the nuclear industry before any new build. (There is some provision for a geological disposal facility which it has been suggested should accept spent fuel waste from the proposed new build.) But after several years of expenditure, approaching £3bn per year, the nuclear liabilities have increased to £44.5bn. Military nuclear liabilities are not included in that estimate, but are of the same order.

Deep geological disposal

The Authority's largest project is the design, construction and maintenance of a deep geological disposal facility. The planning of the facility forms part of the current budget and the government, after a public consultation, has already decided that co-disposal of legacy waste and waste from a new generation of nuclear power stations is desirable and technically possible. The June 2008 framework document *Managing Radioactive Waste Safely*[46] includes the statement:

> *Our policy is that before development consents for new nuclear power stations are granted, the Government will need to be satisfied that effective arrangements exist or will exist to manage and dispose of the waste they will produce. The Government also believes that the balance of ethical considerations does not rule out the option of new nuclear power stations.*

In seems that the government was already satisfied because, six months earlier, it had stated in *Meeting the Energy Challenge* (January 2008[47])

> *Given international experience and the UK's own research, we are confident that a geological disposal facility could be built in such a way as to satisfy the regulators. Safety, security and environmental protection will also be essential in ensuring that there is robust interim storage of waste before the geological disposal facility is developed, commissioned and available for use. Given the ability of interim stores to be maintained in order to hold waste safely and securely if necessary for very long periods (stores currently being constructed for the NDA are designed to last for at least 100 years), or if necessary refurbished or replaced, we are satisfied that it is reasonable to proceed with allowing operators to build new nuclear power stations in advance of a geological disposal facility being available.*

The reader will also by now have recognised the 'robust' style of the government's writer, who used the word 'robust' 35 times in this same document. One of the eight meanings of the word in the Encarta English Dictionary is *'Characterized by firmness and determination and a refusal to make concessions'*. The same determination seems to have been used to eradicate the ethical case that it is not the proposer but our children and their descendents who will pay for the management of the increased waste legacy and any new failure to contain it. Some balance of ethical considerations would be feasible if nuclear reactors were the only way to reduce carbon emissions. They are not.

The number and type of new reactors
In the consultative document *The Future of Nuclear Power*[48] we find

> ... *we cannot be sure of the timing and number of nuclear power stations that might be proposed. However, a scenario considered during the CoRWM process gives an example of the potential impact of replacing the existing nuclear capacity. The CoRWM Inventory report contains reference to a scenario of the construction of 10 new AP1000 power stations with an operating lifetime of 60 years each that would together generate 25% of the UK's electricity. This well documented scenario is used here purely for illustrative purposes ...*

and there has been more recent mention of the AP1000 reactor. The first thing that we need to know about the 'new generation' of 'standardised' designs is that the AP1000 reactor does not exist. On 26 July 2008, Henry Wasserman had this to say

> *The plans for these reactors have not been finalized by the builders themselves, nor have they been approved by the regulators. There is no operating prototype of a Westinghouse AP-1000 from which to draw actual data about how safely these plants might actually operate, what their environmental impact might be, or what they might cost to build or run. In fact, as the NRC's (Nuclear Regulatory Commission's) June 27 letter notes, Westinghouse has been forced to withdraw key technical documents from the regulatory process. The NRC says this means design approval for the AP-1000 might not come until 2012.*[49]

There is more recent evidence that supports the above contention. It is referred to in the sections of this paper on regulation by the nuclear installations inspectorate and on the justification process.

In Britain a licence to build and operate a nuclear reactor is for a particular location with conditions appropriate to that site to be enforced by the Nuclear Installations Inspectorate. The NII is thus empowered to enforce the conditions with the ultimate sanction of immediate shutdown if the management have not already done so. The conditions can be about system integrity, the safe working life of components and structures, system fault prediction and analysis, emergency procedures, back-up power supplies, standby equipment, staffing, training, competence, the safety culture of the organisation, and many other things. It has to be that way because it must not go wrong. We do not have the space in Britain to get out of the way.

The size of the depository
The volume of a London bus is no longer used as a unit of volume for

nuclear waste. The consultative document *The Future of Nuclear Power* (May 2007) quoted the Committee on Radioactive Waste Management as stating that

> *the projected volume of higher activity waste that will arise up to approximately 2120, following decommissioning of existing nuclear facilities, is 478,000 cubic metres (a volume five times greater than that of London's Royal Albert Hall).*[50]

The consultative document continues:

> *Based on the scenario set out above of the construction of 10 new AP1000 power stations with an operating lifetime of 60 years, we can estimate that the volume of higher activity waste (intermediate level waste plus spent fuel) that would be produced would be approximately 40 900 cubic metres – roughly half the volume of London's Royal Albert Hall. However, to understand the impact that waste from new nuclear power stations would have on the size of a repository, it is important to consider the level of the radioactivity of the new waste, as this is a factor in determining how far apart the waste must be placed.*

There follows a long discussion which supports the feasibility of co-disposal and asserts the principle that

> *developers of any new nuclear power stations would have to meet their full share of waste management costs.*

In conclusion it is estimated that the additional quantities of Intermediate Level Waste (ILW) and High Level Waste/Spent Fuel (HLW/SF) would increase the overall footprint (the underground area of excavation in host rock) of a co-located repository by approximately 50% and add £2bn to a projected overall cost of £10bn – both amounts being undiscounted.

Critics of the Nuclear Decommissioning Authority estimates of the depository footprint and costs note that the type of reactor and the burn-up rate are not yet known. At some of the high burn-up rates mentioned the footprint could be as much as three times larger.[51] Matters not yet known or fully evaluated include:

- the fuel burn-up rates allowed in the reactors (stated in megawatt days per tonne of uranium – MWd/t;
- the duration of the cooling off period in spent fuel stores (several decades have been mentioned, even 100 years);[52]
- the contribution of military spent fuel and other waste (excluded from the NDA's remit but now mentioned on the same page 138 of *The Future of Nuclear Power*);

- whether some plutonium is to be treated as waste;
- the detailed design of the depository and the length of time that it is to remain open: this could be more than 100 years if waste from 10 reactors is to be accommodated;
- the construction, operation and maintenance cost of a long life depository;
- the insurance liabilities and any waivers that are to apply;
- the cost of supporting and compensating volunteering communities;
- the cost of anti-terrorism measures, public information and emergency plans at surface and underground sites and transport routes.

Depository site selection

The government's view, after consultation, remains that '*an approach based on voluntarism and partnership is the best means of siting of a geological disposal facility*'. How this may work is set out in Chapter 6 of the DEFRA white paper[53] where it is envisaged that an '*Expression of Interest*' without commitment and with a right of withdrawal may be made by a local authority, a Parish Council, an organisation or a landowner within an area. Local authorities will be expected to measure support for such an interest. Where there is a decision to participate the government proposes the setting up of a *Community Siting Partnership* which will include wider local interest groups to work with the Nuclear Decommissioning Authority's delivery organisation – the recently named *Radioactive Waste Management Directorate* (RWMD).

Chapter 5 of the same white paper, which deals with regulatory agencies and other organisations involved in the planning of waste disposal, describes the processes of '*Strategic Environmental Assessment*', '*Sustainability Appraisal*' and '*Environmental Impact Assessment*'. In August 2008, the Nuclear Decommissioning Authority published a consultative document with the title *A Framework for Sustainability Appraisal and Environmental Assessment for Geological Disposal* which invited responses until 30 November 2008.

At an early stage after *an expression of interest* the British Geological Survey will be asked to apply sub-surface screening criteria in order to eliminate any area that is obviously geologically unsuitable. At Annex B of the white paper criteria for the exclusion or initial inclusion of sites are listed; deep coal, oil and gas fields, for example, are listed for exclusion but areas of evaporite minerals are not. Areas subject to earthquakes, uplift or erosion are not excluded and remain available for later assessment.

Depository design, operation and maintenance
Chapter 4 of the white paper cited above includes a drawing of a *'Generic co-located disposal facility'* which is reproduced below. It is in the style of several similar illustrations published earlier by UK government departments, and the legend is remarkably similar to the published illustration of a Swedish facility. It could be the work of a government draughtsman rather than that of a mining engineer who would hesitate to plan for a five-way junction of slowly converging roadways. Three thousand metre long roadways are envisaged at depths up to 1000m. Three routes of ingress and egress are proposed, one of which is an inclined spiral drift. Separate areas are assigned for high level waste and spent fuel.

There is room for much speculation as to what geological formations at what depths would allow such excavations to remain open for perhaps 100 years, about how much of the 'footprint' may be unusable because of faults and other discontinuities, and how the

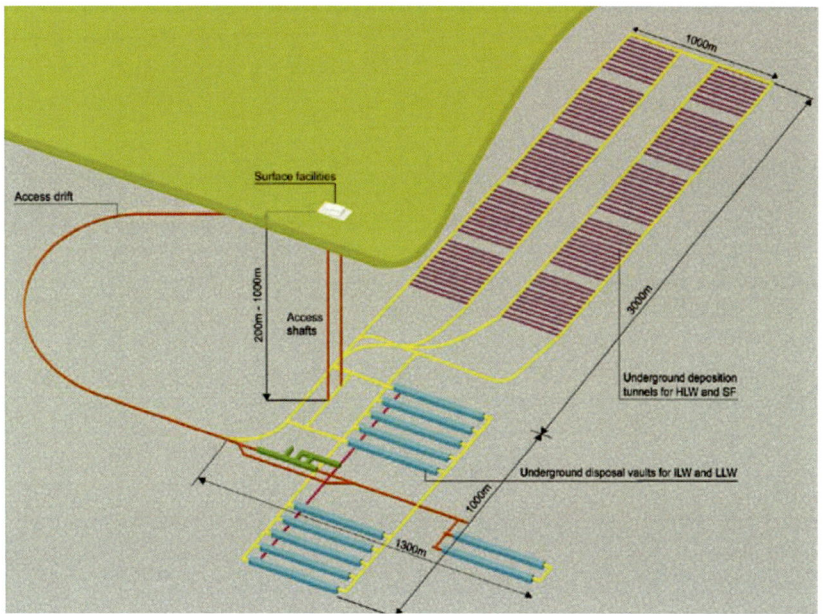

A generic co-located disposal facility

additional heat load of spent fuel can be managed at those depths where the strata temperature is close to that of the human body. Dust control will involve increased humidity and vulnerability to heat stress. The limiting conditions will be those where the effective temperature, in the absence of power supplies for ventilation and cooling, is too high for rescue teams in breathing apparatus to operate for periods long enough to be effective in the circumstances of failed ventilation and air conditioning. The strata itself can be a vast heat sink and co-disposal at shallower depths is less likely to be a problem.

The white paper cited above at page 27 states that, in the course of 2008-9, the Nuclear Decommissioning Authority will undertake early planning for the implementation of a geological disposal facility and that

> *This will include provision for a staged implementation, with clear decision points, that allows design and development, costs, affordability and value for money, safety, and environmental and sustainability impacts to be reviewed at the end of each stage before a decision to move on to the next stage is agreed with government. This planning will be refined and costed as the implementation programme proceeds.*

Interim Storage
The white paper *The Future of Nuclear Power* at page 140 states

> *The provision of interim storage over the life of the plant will be the responsibility of the operator*

with more detail at page 107,

> *In line with the principles on waste management and decommissioning that the Government published in the 2006 Energy Review report, developers would have to provide and pay for flood management after operation has ceased and until any material in interim storage had been removed from the site.*

The possibility of the operator becoming insolvent is dealt with elsewhere in a proposal to create trust funds large enough to meet such a contingency. What is not dealt with in any detail is the vulnerability of on-site storage to terrorist attack. It is in this area that the white paper is least convincing. At page 110 it summarises some of the findings of the Parliamentary Office of Science and Technology report No 222[54] that some existing nuclear power stations were not designed to withstand large aircraft impact but that existing safety and security regimes provide *'some defence'*, and then added:

> *It is important to note, however, that the POST report looked primarily at existing nuclear facilities. Many modern nuclear facilities are designed to withstand the impact of an aircraft. Safety measures can include double layered, reinforced reactor buildings and the strategic siting of protection systems.*

From which we can infer that not all modern nuclear facilities are so designed and that spent fuel stores may be vulnerable.

Total waste management costs and their apportionment

Few, if any, nuclear power projects have been completed to budget and on time and even the current EDF/Areva/Siemens construction at Olkiluoto in Finland is reported to be three years late and 50% over budget. Such history has to be taken into consideration when appraising the government's proposal to charge a 'fixed unit price including a significant risk premium' for new operators for the disposal of waste[55] when the characteristics of the waste, the design of the depository and the cost of its construction and maintenance are not known.

In the 2009 debate in the House of Commons on the second reading of the Energy Bill a new Clause 7 was introduced to require the payment of a *'significant risk premium'* in addition to a *'fixed unit price'* for the disposal of nuclear waste. The risk premium was described as a 'fee' to be decided by the Secretary of State with the approval of the Treasury. The Secretary of State explained

> *'The risk premium should help ensure that the operator bears the risks associated with uncertainty in waste costs. We believe that it will provide the taxpayer with protection against the eventuality that the actual costs of disposal exceed the projected costs.'*

The companies listed as interested in the building of new nuclear power stations (the 'requesting parties') have an interest in the fixing of the 'fixed unit price' and the method of paying it years ahead of the packaging, transporting and deposition of the waste in a geological disposal facility. That MPs were sceptical that the cost of managing the waste was being underestimated could be the reason for the amendment of the Bill to require 'a significant risk premium'. Even that is not enough to allay fear that an unreasonable burden will be placed on the taxpayer in 160 years time. Added to the fact that the cost of building the depository and the cost of the repeated repackaging of the waste are not known, a paragraph in the Department of Energy and Climate Change (DECC) consultation

document on the 'Methodology to Determine a Fixed Unit Price ...'
(DECC March 2010) describes the best contrivance of all.

> *Under the proposals for early transfer, the Fixed Unit Price will be paid many years
> before the Assumed Disposal Date. It is therefore necessary to adjust the payments
> made by the operator to reflect this early payment and this will be done through the
> application of an appropriate discount rate to the Fixed Unit Price to reflect this time
> difference. This discount rate will be determined nearer to the Transfer Date and set
> in relation to the rates of returns available at that time on long-term investment.*

There are several assumptions necessary to believe that this is
anything other than the creation of a detriment with inadequate or
inappropriate compensation. Will the pound have a predictable
value? Will there be banks as we know them? Will there be a
currency? Will there be interest rates better than inflation? Will there
be engineers with the appropriate skills? Will there be government?
Lastly, who will be the 'beneficiary' of the mature investment? One
estimate is that a Fixed Unit Price equal to 17% of the estimated waste
management cost will be sufficient to justify the abuse of a future
environment and our great-great-great-great-grandchildren.

The consultative document *The Future of Nuclear Power* at page 135
offered only the 2003 Nirex estimate of the depository cost of £10bn
(undiscounted) increased by £2bn by a new build programme. If the
depository is to have a footprint not 50% larger, as estimated by Nirex,
but three times larger, as estimated by Richards[56], the apportionment
will be likely to be not 33% but 75% of the planning, construction,
maintenance and decommissioning costs of the depository.
Construction is not expected to be completed until 2045, and
maintenance is estimated to continue for 100 years or more.[57]

The prospective investors in new build who are already committed
to the costs of design assessment will need to know quite soon what
the fixed unit charge and the significant risk premium will be. No
appraisal of the viability of a project can be made otherwise. We have
seen some of the uncertainties in the estimation of future cost to
which can be added concern about the ethics of discounting a liability
where further underestimation as well as any environmental
detriment will be a charge on future generations. More scrutiny will
be needed than that by the Treasury if charges for waste management
and disposal are not to be the biggest scam since the publication of the
'dodgy dossiers' in the prelude to the invasion of Iraq.

Regulation – the Nuclear Installations Inspectorate

Regulatory Effectiveness

The work of the Nuclear Installations Inspectorate and the Nuclear Safety Directorate presented exceptional challenges for any regulatory agency. The extended work load of the Ministry of Defence and the problems of the UK Atomic Energy Authority, the mounting problem of nuclear waste, the repeated failures of the vitrification plant at Sellafield, the proposed and actual privatisations, and the start of decommissioning were more than enough for an organisation that was short of 12 or more inspectors, without the crisis created by the falsification of fuel rod data and the consequent management changes at BNFL.

The Nuclear Installations Inspectorate had anticipated problems of ineffective regulation and published a checklist for our government and other governments on criteria for an effective regulatory agency:

Effective independence
Established regulatory process
Regulatory effectiveness
Adequate inspectorate powers and sanctions
Internal quality assurance and monitoring

The Work of the Nuclear Inspectorate

Several observers have seen a government policy of 'fast track new build' as dangerous, if only because of the pressure placed on the Nuclear Installations Inspectorate. Their response in *Regulatory Strategy for Energy Developments* (HSE June 2006) was realistic and reassuring, in which they described the international validation of existing licensing procedures. No change has yet been proposed in the procedures for granting site specific licences for the building and operation of nuclear installations, and it is important to note that no application for such a licence has yet been made.

We are now two years into Generic Design Assessment (GDA) and it is becoming increasingly clear that the process is stalled for the lack of intelligible design detail. HSE Bulletins have reported delays in the provision of such detail. The Bulletin, dated 16 February 2010, contains the following

The HSE's Nuclear Directorate, the UK's nuclear safety and security regulator, has raised a Regulatory Issue against Westinghouse's AP1000.

> *Westinghouse is proposing to use a new construction methodology for key structures within the 'Nuclear Island', essentially using a sandwich of steel plates filled with concrete, rather than using more conventional reinforced concrete, which is strengthened with internal steel bars.*
>
> *This is new and we need to be reassured that key structures would be sufficiently robust to protect the reactor's safety systems under normal conditions, and also from severe weather and other external hazards, such as physical impacts. In order to get that reassurance, we need to see appropriate evidence to demonstrate the strength and durability of the structures. In essence, we want to be assured that the structure will hold together. The fact that we have issued a Regulatory Issue does not mean that the design is unsafe – ND is still assessing designs on paper, so any safety detriment is still in the design stage. Westinghouse is considering a number of possible solutions, such as further analysis, testing and possible changes to the design, and intend to provide detailed proposals and supporting evidence by the end of October 2010.*

It is some comfort to know that the 'requesting parties', in this case Westinghouse, and not the taxpayer, are paying for these assessments.

The reorganisation of the Inspectorate

There are also concerns about the government's proposed restructuring of the Inspectorate in a Statutory Nuclear Corporation when no defect in their performance has been described. HM Inspectors would cease to be civil servants and would be appointed by an industry linked 'Statutory Corporation'. Industry links are important in any inspectorate but not in the matter of governance. It is far from obvious that regulatory independence and effectiveness will be improved by this process. Changes have already been made to facilitate the recruitment of well qualified inspectors.[58]

Greenpeace also made a significant response to the consultation on this proposed restructuring:

> *History of the proposal*
> *The original Stone Review, which has led to the proposal to restructure the HSE's Nuclear Directorate (ND) and change it to a Nuclear Statutory Corporation (NSC), has never been made public. There was no public consultation on the original review and its aims. That the proposed changes to the ND, which will supposedly make nuclear regulation decision making more open and transparent, are based on a secret review undermines the purpose and understandably leads to public scepticism. The refusal of Government to release the review in full is part of the reason why it is widely believed that the restructuring is aimed primarily at facilitating new nuclear build.*
>
> *While it is accepted that the Government's overall aim is to improve regulation, Greenpeace is concerned that this is in a context where de-regulation, light-touch*

regulation, or even self-regulation is seen as an improvement by the Government. For the avoidance of doubt, Greenpeace is absolutely clear that regulation of the nuclear industry must be independent, transparent, and thorough. The potential dangers of nuclear power are manifold and accidents catastrophic. There must be no short cuts with nuclear regulation.

There is a contradiction between the Government's policy aim of facilitating nuclear new build and improving resources for regulation which lies at the heart of this consultation. If implemented and taken together with other measures being proposed by Government, the restructuring will not lead to real autonomy for nuclear regulators as it does not provide the necessary distance between those Government departments promoting nuclear power and the regulator.

We note the consultation document states (2.10) the review 'has made a number of recommendations designed to address the ND's immediate (new build) and longer-term needs, and which reflected emerging views within the Government and across the nuclear industry'. We understand the first meeting of a transition committee on the ND becoming the NSC, set up by the HSE, will meet at the end of September. That arrangements are being made on this before the consultation is finished, a Ministerial decision made and prior to Parliamentary scrutiny, is evidence of the pre-determined outcome of the proposal and casts doubt on the openness of this consultation.

Greenpeace recognises the need to encourage new employees to the ND, and retain present employees, to deal with existing nuclear installations. That this may be an unintended consequence of this proposal does not make its basis any more acceptable. The consultation does not offer other means by which the recruitment and retention of staff to deal with existing installations can be achieved.

We note that this consultation was issued before the entering into force of the Council Directive establishing a Community Framework for the nuclear safety of nuclear installations. We do not think that the proposed structure meets the requirements of Article 5.

Statutory justification

Paragraph 5.22 of the White Paper *Managing Radioactive Waste Safely A Framework for Implementing Geological Disposal*[59] explains the legal process of justification in its application to geological disposal of nuclear waste.

European legislation (Ref. 26) requires that any new practice involving ionising radiation initiated on or after 13 May 2000 needs a justification decision from the Member State that the benefits of the practice outweigh any detriment to health that might be caused by exposure to radiation. However, guidance from the International Commission on Radiological Protection (ICRP) (Ref. 37) and Defra (Ref. 38) on behalf of the Justifying Authorities states that waste management and disposal operations are an integral part of the practice that generates the waste and it is inappropriate to regard them as free-standing practices that require their own justification.

The Justification of Practices Involving Radiation Regulations 2004 (SI 2004/1769) came into force on 2 August 2004.

If the guidance quoted is followed it seems that a justification application by the Nuclear Decommissioning Authority need not be made for geological disposal of legacy waste, and that arguments for and against justification could be ruled *ultra vires* in any planning inquiry. But it is already conceded that the government will justify its own 'facilitative action' in support of new build (p.176 of *The Future of Nuclear Power*) and that must include the management and disposal of new build waste.

The previous government's planning reforms could have had the effect of making an '*Infrastructure Planning Commission*' responsible for deciding, after consultation, on an application for a development consent.[60] The Coalition Government intends to abolish the Commission and publish new proposals.

The guidance quoted above providing relief from justification will not apply to the building of power stations, which are radically different from those already the subject of a consent, and the Nuclear Industry Association (NIA) has already made a justification application for types of reactors offered by its members. Part 1 of the application is a 107 page document accessible on the DEFRA web site, http://www.defra.gov.uk. It concludes that the benefits outweigh the detriments and that the practices will be justified. Part 2, obtained as an e-mail attachment, has 122 pages and descriptions of reactor designs without specification of detail such as fuel regimes.

In 2008, the Justifying Authority was in the process of assessing whether sufficient information has been provided in the Nuclear Industry Association's application and, where necessary, requesting additional information.[61] Whether the Department of Energy and Climate Change (DECC) should be the sole justifying authority when the protection of the environment, unlike energy policy, is a devolved matter, remains a matter for debate.

The Secretary of State in the last government drafted a Justification Statutory Instrument and published it for consultation. I and many others, mainly specialists, responded to the consultation. The process was halted by the general election.

The new Secretary of State for Energy and Climate Change is the Rt

Hon Chris Huhne, a Liberal Democrat MP, whose party policy favoured renewable energy but not nuclear power generation. The negotiations on the coalition agreement which preceded the formation of the present government took account of the fact that Conservative Party policy was favourable to nuclear electricity, subject to the proviso that there must be no subsidy by the taxpayers, and allowed for Liberal Democrat MPs to vote independently of government policy. It remains to be seen whether Chris Huhne will publish a draft justification instrument similar to that of his predecessor. Perhaps some other minister will be asked to deal with the matter.

My personal response to the original draft justification instrument is reproduced in part in the paragraphs below. With the possible exception of the expected revised appraisal of sustainability, nothing has occurred to invalidate my response dated 18 February 2010[62] and, if necessary, I will revise it in response to any new justification consultation.

I am an adviser to the Welsh Anti-Nuclear Alliance whose members include Greenpeace and Friends of the Earth, which organisations made separate responses to the earlier consultation with which I largely agree.

I advised that the Secretary of State designated to make the decision on justification should think long and hard about making the Statutory Instrument as drafted justifying the building of new nuclear power stations. It could be a decision that will haunt him for the rest of his political career. For the rest of us and for posterity it will be one that lasts long into the future, making his reputation and that of the government hostage to many foreseeable detriments. The collapse of the proposals by the Nuclear Industry Association would be one of the least embarrassing outcomes. Much worse would be a nuclear mishap bringing the whole nuclear power project to a second standstill.

The arguments and reasons written for the Secretary of State in Volumes 1 and 2 of the consultation are unconvincing, even for advocates of the processes described. He will do well to read them wearing a political hat and with just enough information about disasters in high risk industries to prompt some scepticism. He may remember the words of the former HM Chief Inspector of Nuclear Installations, after his 1986 visit to Chernobyl, that 'things will never be the same again'.

The nuclear industry faded because scientists, engineers, entrepreneurs, investors, politicians and the public found the risks, hitherto poorly described but demonstrated at Windscale, Three Mile

Island, Sellafield, Dounreay and Chernobyl, intolerable. Could it really be possible to have to evacuate a whole town for many years or to have food supplies as far away as Wales jeopardised until even now? Was it possible that those in charge of a reactor would deny the need for evacuation or hope to keep secret the extent of the danger? The possibility of much worse outcomes, such as making large tracts of Britain uninhabitable by mishap or terrorism, is effectively denied in these volumes without even providing the names of the those who find the risks 'negligible'.[63,64]

Detriments of nuclear understated and evaded
It seems that the detriments of nuclear power generation are carefully understated in this consultation, even evaded. Why is it 'inappropriate' to estimate the number of additional cancer deaths attributable to the nuclear industry?[65] Estimates of an increased number of birth defects in humans and other animals could also be made, but the industry and its promoters desist from creating anxiety for young mothers and fathers. The incidence could well be low if the doses planned for workers and the public are achieved. They have not always been achieved, and any estimates based on good management must be adjusted to take account not only of accidental losses of containment but also those created by hostile acts.

Sustainability appraisals and environmental impact assessments
Strategic environmental assessments (SEAs) of proposed plans are required by EC Directive 2001/42/EC and Sustainability Appraisals are required in England in relation to the aims of sustainable development under the Planning and Compulsory Purchase Act 2004. There are similar requirements in the devolved administrations. The government expects the Nuclear Decommissioning Authority to undertake Sustainability Appraisal incorporating a strategic environmental assessment and the Authority, in August 2008, published a consultative document on *A Framework for Sustainability Appraisal and Environmental Assessment for Geological Disposal.*

The government is also committed to such appraisals, including justification, in relation to its 'facilitative action' in support of the building of new nuclear power stations[66], but one detects an emphasis on a quick result rather than a concern for sustainability. In a paragraph on strategic environmental assessment it concludes: '*This* would limit the need to consider such high-level environmental impacts of nuclear power stations during the planning process'. It is paradoxical that the government as facilitator will likely justify a process to itself. Such conflict of interest, if not resolved by better process, will leave a sceptical public unconvinced.

Sustainability is about doing those things that one can keep on doing without harm. It is about leaving the planet at least as good as one found it. The fancy word for it is intergenerational equity – not abusing the children of the future. It is also about resource depletion and, as E F Schumacher made clear 40 years ago, finite resources are exhaustible.

Uranium and Thorium are no less exhaustible than fossil fuel and, ultimately, we have to rely on renewable energy sources.

In the documents quoted in this paper the government's disregard of renewable energy and energy conservation is unmistakable. When discussing alternatives to nuclear energy the discussion is largely about wind turbines, which can provide only a fraction of our needs. In a search for the word 'tidal' in the 2008 *'Meeting the Energy Challenge'*, the mentions are largely what respondents to consultations had to say. In contrast the white paper with the same title, published nine months earlier, regarded tidal energy as a significant resource.

The practicability of the successful management of the nuclear waste generated in the last 60 years has never been demonstrated. The record is one of failure, underestimation, misinformation, procrastination and neglect. The resources for the clean up have yet to be found, and it is highly unlikely that they will be found in our lifetimes. In other words, we will ask our children to pay. There are powerful reasons for believing that problems that will last for hundreds of thousands of years, requiring better government than has been demonstrated to date, will never be solved, and that our legacy to future generations will be seen simply as an abuse of them and the planet. Perhaps they will have a name for us; we who messed up for 60 years and even then thought of making more mess.

Will the benefits of low carbon nuclear power come too late?
No new nuclear power station is likely to produce electricity before 2018[67] and, as the government has already conceded, using only mid range estimates of CO_2 footprint and stating no CO_2 emission cost from the excavation, transport and milling of uranium ore, a new build programme could have only negative effects on atmospheric carbon dioxide until 2023.[68] If, as is suggested by the New Economics

Foundation[69], what is done to reduce CO_2 emissions in the next 100 months is critical to preventing irreversible climate change by the loss of surface ice and the release of methane by the thawing of permafrost, then effecting no change until 2026 will amount to failure.

The year 2015 is not far away. If the urgency is to replace 20GW of generating capacity in the next seven years one thing is clear: there is not enough time to build nuclear stations. To mitigate climate change and to maintain a safe reserve of generating capacity we need action with quick results.

Decentralised electricity is the first step away from massive stations with cooling towers to get rid of waste heat at sites remote from populations with long transmission lines and consequent transmission losses. Such stations convert only 33% of the energy input into useful electricity.

The quickest way to build new stations is to build smaller stations using conventional fuels, probably gas, as a short term measure. They can be built near centres of population and industry to provide combined heat and power (CHP). Modern gas stations using waste heat for space heating can be 70% efficient.

Demand management can further reduce the need for generating capacity by more efficient appliances, better heat insulation and local, or 'micro' combined heat and power. There are large savings to be made by combining the heating of larger premises with local generation of electricity connected to the grid.

False information that supported the policy change

No new build was proposed for 20 years after Chernobyl. What changed? Was it the availability of new, proven, safer, simpler, cheaper technology? Such claims were made and they are appraised below. Between 2002 and 2004 UK government policy changed. The reason for the change was not made clear. It was not climate change or the need for low carbon electricity, nor was it security of electricity supply. Those issues had been dealt with adequately in the 2002 *Energy Review*.[70]

The Nuclear Industry Association couldn't believe its luck. There was to be fast tracked progress for 'new build' on all fronts, and no more troublesome public inquiries. More dodgy dossiers appeared and the misleading claims persist.

> *'The (advanced Generation III+) AP1000 is a 1154 MW nuclear power plant that uses the forces of nature and simplicity of design to enhance plant safety and operations and reduce construction costs.'*[71]

This is simply not true because the AP1000 does not exist. Nor was its claimed precursor, the AP600, ever built. The Westinghouse website, like other publications, eventually makes it clear that the AP1000 is a concept, just a design. No one in the industry is likely to have been misled, but was any member of the public or Member of Parliament? My MP in a public meeting put it to me that this time we would be building improved, proven designs. Perhaps he had read

> *True – the candidate designs for new build in the UK do not originate here but this is a big plus for the potential developers who want the confidence that they will be building a proven international design, already built elsewhere in the world. ... Research has shown that 70-80% of a new plant can come from UK companies, and we in Westinghouse are already working closely with major potential UK suppliers such as Rolls Royce, BAE Systems, Donsan Babcock and Sheffield Forgemasters, as part of our Buy Where We Build policy for the AP1000 reactor.*[72]

When I wrote to the journal *ProFile* asking the writer, Mr Adrian Bull of Preston, where the AP1000 had been built elsewhere in the world, there was no reply.

Now we have to ask how is the 'proven international design' standing up to the government's Generic Design Assessment (GDA) procedure by the Nuclear Installations Inspectorate and the Environment Agency. The Nuclear Directorate's *Newsletter* dated September 2009 on the now two-years-old GDA programme states of the two designs submitted for assessment (the AP1000 and the Areva-EDF UK-EPR) that

> *The present position is that neither design is complete, which makes our assessment more difficult. The greater the shortfall in the content and clarity of the information submitted by the Requesting Parties, the more difficult our assessment becomes, with a greater chance of TQs (Technical Queries) being elevated to become more serious 'regulatory observations' or 'regulatory issues'. This in turn is likely to lead to more areas being excluded from the GDA confirmation (using what are presently called 'exclusions'), and the less meaningful the GDA confirmation will become as a means of providing design assurance.*

So the AP1000 exists as an incomplete design.

I thought it important that Members of Parliament dealing with the government's *Draft National Policy Statement for Nuclear Power Generation (EN-6)*[73] should know of the claimed existence of a non-existent reactor and its promoter's inability to produce an intelligible design. I offered to present evidence to the Commons Energy and Climate Change Committee and prepared to quote also and comment on a major design fault found by the joint regulators in the safety system of the EPR/Areva reactor – the second of the two surviving designs of reactors submitted for generic design assessment.

A reactor safety system has to be designed to deal with failures of all kinds including failures of the control system or loss of access to the control system. The UK Nuclear Installations Inspectorate and other regulators in France and Finland have found that the safety system of the EPR is not independent of the control system, in particular that it requires some functionality of the control system to control the reactor in extreme conditions.[74]

It is odd that this should be found when a consent to build in Finland has already been granted. The design has been described as similar to reactors working in France and one wonders where else this design defect may exist. It is also odd that a reactor design is still being assessed at the construction stage.

The Clerk to the Committee regretted that time limits imposed by the government left no time for an appearance. He undertook to present my paper on geological disposal to the Committee members.[75] One hopes that others had the opportunity to discuss with the MPs the building elsewhere of an incompletely appraised reactor.

Renewable energy

Energy from renewable sources has the greatest potential for reducing CO_2 emissions, and it is regrettable that it was not promoted better before our fossil fuel reserves were abandoned or approached exhaustion. The change to renewable energy will have to be gradual because of the infrastructure changes that will be needed, notably in the grid and in load management. The energy sources available are vast and inexhaustible for as long as there is a sun, a hot core to the planet, and life on earth. Compared with nuclear power the technologies are benign. They include

- Hydro-electricity from tides, using tidal current generators or tidal barrages
- Hydro-electricity from waves
- Hydro-electricity from rivers
- Wind generators
- Solar heating direct or photovoltaic panels on buildings or unused land, eg motorway embankments
- Geothermal energy from hot rock
- Geothermal energy using heat pumps

- Biomass grown on marginal land as vehicle fuel or as fuel for space heating or electricity generation
- Gas from small scale waste retorts or landfill sites.

On the proposed Infrastructure Planning Commission and on local planning inquiries being required to exclude matters of national policy from their considerations, a letter to *The Guardian* on 23 May 2010 summed up the argument very well.

> *If the government builds a nuclear power station on the site of London's derelict Battersea power station then the rest of the country will know that these stations are completely safe. The new streamlined planning system should take care of any local opposition.*

Spending billions more on new build will inevitably impede and distract from the investment that we need to make in several forms of renewable energy, particularly tidal energy. Energy policy impacts directly on the size and cost of a geological disposal facility and who pays for it. The reconstituted Committee on Radioactive Waste Management[76] has commented so far that

> *At present, it is uncertain whether the appropriate combination (or combinations) of community and site can be found in this country. This uncertainty applies to existing and committed highly active waste (HAW), as well as to new build HAW, and is likely to persist for many years.*[77]

Nuclear is not the only low-carbon form of energy
That the government was considering the building of a 'fleet' of 10 non-existent reactors of inadequate design is not the only cause for alarm.[78] The last government based its draft decision on justification by comparing the detriments of nuclear power generation with the detriments of not taking action on climate change by investing in low carbon forms of energy such as nuclear.[79] But low carbon energy is available from many sources, none of which involve radioactive waste or present terrorists with such opportunities for havoc. The option of large scale renewable energy has been understated in recent consultations. It was considered very favourably in the 2002 *Energy Review*[80] (and in the associated white paper). That this option was not pursued fully eight years ago leaves us now with options most of which have long lead times but none quite as long as nuclear.

When writing in 2006, I quoted that the German engineering group Bosch had identified 100 possible locations around Europe for tidal generators with capacity equal to 100 nuclear power stations[81] – a finding similar to that mentioned earlier as reported by the Offshore Valuation Group.[82] I speculated in 2006 that a gap in generating capacity may require the building of more gas stations. That view is now shared by other energy specialists. Such a compromise comes with the assurance that bio-gas is renewable, that natural gas remains abundant, and that, with local combined heat and power generation, gas generators will have a carbon footprint comparable with any new nuclear for the next two critical decades.

Ethical issues
Equity – international and intergenerational

The Chairman of the original Committee on Radioactive Waste Management, Professor Gordon MacKerron, (the committee was reconstituted in 2007) submitted evidence[83] to the Parliamentary Committee on Energy and Climate Change in January 2010, in collaboration with Greenpeace, on ethical issues of waste management. He and Greenpeace make clear the difference between geological disposal as a procedure of least harm for dealing with legacy waste after 60 years of failure, and such disposal as an 'effective arrangement' which we can be sure will exist. It is my opinion, as a mining engineer, that there will be problems in finding geologically stable formations free from existing and future faults that can be conduits for liquids and gas. Containment for the periods of time needed for the protection of those parts of the biosphere on which human life depends, for example, for safe water supplies, cannot be assured.

Professor MacKerron's evidence included that the government's treatment of ethical issues was inadequate, and that the location of high burn-up waste in surface stores, for perhaps 160 years, will diminish any support for new build in the communities involved.

One of the ethical principles underlying consultation and informed consent is that risk bearers should be involved when decisions on risks are taken. It is inequitable that risk takers should benefit if others who do not benefit suffer some detriment. Thus there is an international obligation particularly to those who do not have a supply of electricity.

When those who may suffer a polluted environment and radiological injury are not yet born, justification is simply not possible. In *The Ethics of Environmental Concern*[84] Professor Robin Attfield agrees with R and V Routley that 'there is the same obligation to future people as to the present' (even for 30 000 generations, he adds, for which a discounted financial provision is no remedy) and concludes, with the support of many others,[85] that 'almost any serious decision procedure for the assessment of risk supports the anti-nuclear case'.

Conclusions

'Without cost to the taxpayer' means that every government 'facilitation' of nuclear power has to be costed and, if found significant, removed. Thus, insurance waivers have to be removed and the pretence of paying for future waste management by providing only 17% of the apparently underestimated cost has to be abandoned. Also stopped must be any Nuclear Decommissioning Authority expenditure other than that on legacy waste.

Producing spent fuel waste 'too hot to handle' for many years, and for others to deal with in 160 years time, if they can, is just one of many unresolved issues before the Coalition Government which is probably debating the justification for new build in cabinet.

There are signs that a reorganised Nuclear Installations Inspectorate (NII) is being given a new culture in which it will lack the necessary independence. The NII exists to protect the public and, with the Environment Agency, the environment. They do not exist to promote government energy policies. A sign of the required independence of government will be that the NII makes plain to government by reiteration of its own reports that the consistent safe management of nuclear waste has not been demonstrated in the last 60 years.

Devolved governments have also to be involved in Statutory Justification.

The designated Secretary of State is urged to desist from making a renewed Justification Statutory Instrument and, instead, to exercise the option of holding a public inquiry. This paper has set out to demonstrate the many issues of fact and opinion that remain to be examined and, if possible, resolved. Here is a list:

- To question why the Royal Commission on Environmental Pollution is being abolished instead of being allowed to appraise and comment on current policies.
- To examine the claim that modern reactors of proven safe design exist when the Nuclear Installations Inspectorate has reported that only incomplete and inadequate designs have been submitted in the last three years.
- To note that no application to build and manage a nuclear power

station has yet been made and to seek and report an explanation.

- To examine the last government's claim that it had solutions for the safe management of nuclear waste.
- To note that the recommendation by the first Committee on Radioactive Waste Management on geological disposal of legacy waste was for reasons of 'least harm' and was never offered as a 'solution' or as a method for co-disposal of extremely highly active spent fuel.
- To find the reasons why 'flawed' and 'unlawful' consultations were made, and whether they are still being made, for example, on the reorganisation of the nuclear regulators.
- To discover why a sudden policy change was made in 2005 to 'facilitate' nuclear new build and to describe the extent to which the cabinet was involved.
- To hear and examine the Coalition Government's proposals for replacing the Infrastructure Planning Commission.
- To name and produce for cross-examination the authors of the statements that the risks from the nuclear industry are negligible.
- To examine witnesses who were authors of the statements that spent fuel stores and highly active liquor stores could be made safe and secure from attack by terrorists, and to require them to explain how that may be achieved.
- To find why the Radiation (Emergency Planning and Public Information) Regulations 2002 have not been implemented to deal with the foreseeable effects of attacks on nuclear installations by terrorist organisations.
- To investigate the legality of making spent fuel nuclear waste available for use as ammunition by the military.
- To examine the future security of electricity supply by comparing all the available methods of low carbon electricity generation.
- To question those responsible for the last government's dismissal, in a single sentence, of the ethical issues of nuclear waste production[86], especially highly active spent fuel waste.

'The Government also believes that the balance of ethical considerations does not rule out the option of new nuclear power stations.'

Christopher Gifford
October 2010

References

1 *The Energy Review* Dept of Trade and Industry (DTI) London November 2005 followed by *The Energy Challenge: Energy Review Report 2006* Cmnd 6887 DTI London SWIA.

2 *Guardian 1 June 2010:* 'Huhne warns of £4bn black hole in nuclear budget' (and 'flagged the crisis up to the cabinet'.)

3 Margaret Beckett *Hansard* House of Commons debates for 18 October 2001.

4 www.offshorevaluation.org

5 The Coalition: Our programme for Government. http://programmeforgovernment.hmg.gov.uk/files/2010/05/coalition-programme.pdf

6 *Nuclear Power and the Environment (also known as the Flowers report)* Sixth report of the Royal Commission on Environmental Pollution Cmnd 618 HMSO 1976.

7 Oxford English Dictionary 'a. 2a. having sound judgement; sensible, sane, = RATIONAL a 1b. Also, not expecting too much.' Encarta Dict. UK 'not expecting more than is possible'.

8 *Managing Radioactive Waste Safety Proposals for developing a policy for managing solid radioactive waste in the UK* Page 15 Dept of the Environment DEFRA The Scottish Executive The National Assembly for Wales Published by DEFRA 123, Victoria St London SW1E 6DE 25 June 2007 Ref PB12646

9 *Managing our Radioactive Waste Safely* CoRWM's Recommendations to Government Final Report November 2006 CoRWM, Ashdown House, 123 Victoria Street, London SW1E 6DE.

10 The new committee appointed in October 2007 with revised terms of reference will provide independent scrutiny and advice to UK government and the devolved administrations on the long-term radioactive waste management programme, including storage and disposal.

11 CEGB officials at the Hinkley Point Public Inquiry (Cardiff hearing) objected to being asked to confirm this exchange on the grounds that Lord Hinton was dead. The witness, Dr John Cox, responded with 'But he was alive when he said it'.

12 *Sixty Years of Nuclear History: Britain's Hidden Agenda* John Carpenter Publishing, Charlbury Oxon UK OX7 3PQ 1999 £12.

13 *Profile*, the Journal of the Trade Union Prospect, February 2006.

14 HSE Nuclear Safety Directorate *Intermediate Level Waste Storage at Sellafield 1999* HSE Books Sudbury Suffolk CO10 6FS

15 A Department of Health survey in 1990 found plutonium in the teeth of every teenage child examined in Britain. The survey of 3300 adolescents showed minute traces of plutonium in amounts correlating with the distance from Sellafield. O'Donnell et al, *Variations in the concentration of plutonium, strontium 90, and total alpha-emitters in human teeth collected within the British Isles.*

16 *The Energy Review* A Performance and Innovation Unit Report Cabinet Office The Mall London SW1A 2WH February 2002

17 *The Energy Review* Dept of Trade and Industry (DTI) London November 2005 (op cit) followed by *The Energy Challenge: Energy Review Report 2006* Cmd 6887 DTI London SWIA (op cit).

18 *Nuclear Reactors – Do we Need More?* The paper is available as an A5 booklet of 33pp with 61 source references price £2.00 from Spokesman Books Nottingham Tel 0115 9708318

19 *The Health and Safety Risks and Regulatory Strategy Related to Energy Developments* An expert report by the Health and Safety Executive contributing to the government's Energy Review June 2006 HSE Books Sudbury Suffolk CO10 2WA Tel 01787 313995

20 *Planning for a Sustainable Future* Cmnd 7120 Communities and Local Government Eland House London SW1E 5DU May 2007

21 *Meeting the Energy Challenge – A White Paper on Energy* DBERR May 2007 Cmnd 7124 TSO Norwich NR3 1GN Tel 0870 240 3701

22 *The Future of Nuclear Power – the role of nuclear power in a low carbon economy* Consultative Document May 2007 The Department of Trade and Industry 1 Victoria St London SW1H 0ET

23 See the HSE website for the report of the independent GDA Process Review Board: Hughes, D; Raine, J; Whittle, B; and Woodward, P; March 2008.

24 *The Future of Nuclear Power* op cit

25 Current Approaches to Human Factors in the Mines and Quarries Inspectorate; Evidence submitted by C Gifford to the Hinkley Point 'C' Nuclear Power listed as Public Inquiry Document S2327 and the Transcript of Proceedings of the Inquiry on Day 59, 31 January 1989, pp96-107 on the author's cross-examination of Mr J D Rimington, Director General of the Health and Safety Executive. Social Studies Library The University of Wales, Colum Road, Cardiff UK

26 This topic, on which at least two Fellows of the Royal Society have stated that opinion and which is not yet resolved, was described in some detail in *Nuclear Reactors- Do we Need More?* op cit. Page 10.

27 Gifford, C, *Geological Disposal of Nuclear Waste;* Paper presented for discussion at a meeting of the Institute of Materials, Minerals and Mining on September 16 2000 at Cardiff University Department of Earth Sciences and later posted on the website of the South Wales Institute of Engineers Educational Trust http://www.swieet2007.org.uk/files

28 The Radiation (Emergency Preparedness and Public Information) Regulations 2001 SI 2975

29 DECC November 2009 Presented to Parliament pursuant to section 5(9b) of the Planning Act 2008.

30 Draft National Policy Statement for Nuclear Power Generation (EN-6)

Dept of Energy and Climate Change Nov 2009

31 The law requiring such information is REPPIR (The Radiation (Emergency Preparedness and Public Information) Regulations 2001). It remains disregarded.

32 *The Future of Nuclear Power* op cit. page 9

33 *Managing Radioactive Waste Safely – A Framework for Implementing Geological Disposal: A White Paper* by DEFRA, DBERR and the devolved administrations for Wales and Northern Ireland Cmd 7386 Page 92 June 2008 TSO Norwich NR3 1GN Tel 0870 240 3701

34 Quoted in *Too Hot to Handle*, a paper by Hugh Richards – submitted by the Welsh Anti-nuclear Alliance (WANA) to the Welsh Assembly Government in response to the consultation on waste management and discussed with Jane Davidson, Minister for the Environment, on 23.6.08.

35 Ibid

36 *Managing Radioactive Waste Safety Proposals for developing a policy for managing solid radioactive waste in the UK* Page 15 Dept of the Environment DEFRA The Scottish Executive The National Assembly for Wales Published by DEFRA 123, Victoria St London SW1E 6DE 25 June 2007 Ref PB12646

37 Quoted in *Too Hot to Handle*, a paper by Hugh Richards – op cit

38 Draft National Policy Statement for Nuclear Power Generation (EN-6) Dept of Energy and Climate Change Nov 2009 page 25

39 *Managing Radioactive Waste Safety Proposals for developing a policy for managing solid radioactive waste in the UK* September 2001 Dept of the Environment DEFRA The Scottish Executive The National Assembly for Wales Published by DEFRA 123, Victoria St London SW1E 6DE September 2001 Ref. PB5957

40 *Safety Audit of Dounreay* HSE Nuclear Safety Directorate 1998

41 David Gibson's presentation at the Department of Earth Sciences, Cardiff University, on 20 February 2007 was entitled *Dounreay Shaft Isolation – Advanced Rock Grouting in the Nuclear Waste environment.* It is not known whether or not publication is intended.

42 *Managing Radioactive Waste Safely – A Framework for Implementing Geological Disposal: A White Paper* by DEFRA, DBERR and the devolved administrations for Wales and Northern Ireland Cmd 7386 June 2008 TSO Norwich NR3 1GN Tel 0870 240 3701

43 The Health Hazards of Depleted Uranium Munitions: Part 1. The Royal Society May 2001.

44 The Environment Agency: Technical Report P3-088/TR 27pp with 42 references. November 2001. Alan Martin Associates.

45 *Managing Radioactive Waste Safely: DEFRA and the Devolved Administrations* September 2001

46 *Managing Radioactive Waste Safely – A Framework for Implementing Geological*

Disposal: A White Paper by DEFRA, DBERR and the devolved administrations for Wales and Northern Ireland Cmd 7386 June 2008 TSO Norwich NR3 1GN Tel 0870 240 3701

47 *Meeting the Energy Challenge: A White Paper* on Energy January 2008 DBERR

48 *The Future of Nuclear Power – the role of nuclear power in a low carbon economy* Consultative Document May 2007 Page 131 The Department of Trade and Industry 1 Victoria St London SW1H 0ET

49 Article by Henry Wasserman in CommonDreams.org 26 July 2008

50 *The Future of Nuclear Power – the role of nuclear power in a low carbon economy* Consultative Document May 2007 Page 132 The Department of Trade and Industry 1 Victoria St London SW1H 0ET

51 *Burying the Truth* and *Too Hot to Handle*, papers by Hugh Richards – submitted by the Welsh Anti-nuclear Alliance (WANA) to the Welsh Assembly Government in response to the consultation on waste management and discussed with Jane Davidson, Minister for the Environment, on 23.6.08.

52 *The Future of Nuclear Power – the role of nuclear power in a low carbon economy* Consultative Document May 2007 Page 137 The Department of Trade and Industry 1 Victoria St London SW1H 0ET

53 *Managing Radioactive Waste Safely – A Framework for Implementing Geological Disposal: A White Paper* by DEFRA, DBERR and the devolved administrations for Wales and Northern Ireland Cmd 7386 June 2008 TSO Norwich NR3 1GN Tel 0870 240 3701

54 *Assessing the Risks of Terrorist Attacks on Nuclear Facilities* July 2004 available on the Parliament website
www.parliament.uk/documents/upload/POSTpr222.pdf

55 The Minister, Malcolm Wicks, in the House of Commons debate on the second reading of the Energy Bill. Hansard Debates for 30 April 2008.

56 *Burying the Truth* and *Too Hot to Handle*, papers by Hugh Richards – submitted by the Welsh Anti-nuclear Alliance (WANA) to the Welsh Assembly Government in response to the consultation on waste management and discussed with Jane Davidson, Minister for the Environment, on 23.6.08.

57 *The Future of Nuclear Power – the role of nuclear power in a low carbon economy* Consultative Document May 2007 Page 137 The Department of Trade and Industry 1 Victoria St London SW1H 0ET

58 Extract from the author's response to the consultation on the proposed Justification of Nuclear New Build. DECC website.

59 *Managing Radioactive Waste Safely – A Framework for Implementing Geological Disposal: A White Paper* by DEFRA, DBERR and the devolved administrations for Wales and Northern Ireland Cmd 7386 June 2008 TSO Norwich NR3 1GN Tel 0870 240 3701

60 Department for Communities and Local Government, *Planning for a Sustainable*

Future: White Paper, May 2007 www.communities.gov.uk/publications

61 Stated in an e-mail from Simon.Dilks@berr.gsi.gov.uk August 2008 to the Welsh Anti-Nuclear Alliance who were already of the view that NIA's information on spent fuel was inadequate and misleading as to its impact on the size of a depository.

62 I authorised the publication of my response to the Justification consultation on the DECC website.

63 Volume 2; paras 9-35, 9-36 and 10-6

64 In response to this finding by the consultants for the Association of Nuclear Free Local Authorities the then Chief Inspector of Nuclear Installations, Dr Sam Harbison, commented only that it was a low probability event of high outcome.

65 Paragraph 3.83 of Volume 2 the consultative document, op cit.

66 See page 176 of *The Future of Nuclear Power – the role of nuclear power in a low carbon economy* Consultative Document May 2007 The Department of Trade and Industry 1 Victoria St London SW1H 0ET

67 Written statement 15.7.10 by the Minister of State for Energy that the government is launching a re-consultation in the autumn on the draft energy National Policy Statements due to the changes which have been made to the Appraisal of Sustainability for the Overarching Energy National Policy Statement.

68 The chart at page 93 of *The Future of Nuclear Power 2007* shows no net benefit until 2023.

69 Andrew Simms, Policy Director of the New Economics Foundation, proposing a 'Green New Deal' in *The Guardian* 1 August 2008

70 *The Energy Review* A Performance and Innovation Unit Report Cabinet Office Feb 2002

71 From the web site of the Westinghouse Corporation 10 February 2010

72 *ProFile* – the journal of the trade union Prospect, September 2009

73 DECC November 2009 Presented to Parliament pursuant to section 5(9b) of the Planning Act 2008. At page 25 of this 300 page document the government concludes that the Infrastructure Planning Commission need not consider the management and disposal of nuclear waste because it is satisfied that effective arrangements will exist.

74 HSE Joint Regulatory Position Statement on the EPR Pressurised Water Reactor January 2010

75 http://www.swieet2007.org.uk/files

76 The new committee appointed in October 2007 with revised terms of reference will provide independent scrutiny and advice to UK government and the devolved administrations on the long-term radioactive waste management programme, including storage and disposal.

77 CoRWM response to government on the draft National Policy Statement

for Energy Infrastructure, page 2, para 12 March 2010.

78 See paragraph 4-42 of Volume 2 of the consultative document on the justification decision. DECC November 2009.

79 Paragraph 4.145 of Vol 2. op cit.

80 *The Energy Review* A Performance and Innovation Unit Report Cabinet Office Feb 2002.

81 Franz Fehrenbach, Chief Executive of Bosch, reported by David Gow in *The Guardian* 28 April 2006. Quoted in *Nuclear Reactors: Do we need more?* Christopher Gifford Spokesman Books, Russell House, Nottingham NG6 0BT

82 www.offshorevaluation.org

83 A Key Nuclear Question that Government Shrugs off as a Waste of Time. Gordon MacKerron, Parliamentary Brief 7 January 2010
http://www.parliamentarybrief.com/articles/1/new/mag/77/1038/a-key-nuclear.html

84 *The Ethics of Environmental Concern.* Robin Attfield, Second Edition 1991, University of Georgia Press, Athens, Georgia, and London. p94.

85 Including Kavka, G; Rawls, J; Routley, R and V; Goodin, R; Lippit, V D; and Hamada, K.

86 *Managing Radioactive Waste Safety* June 2008 op cit.